FARMALL REGULAR

PHOTO ARCHIVE

FARMALL REGULAR

PHOTO ARCHIVE

Photographs from the
McCormick-International Harvester Company Collection

Edited with introduction by
P. A. Letourneau

Iconografix
Photo Archive Series

Iconografix
P.O. Box 18433
Minneapolis, Minnesota 55418 USA

The photographs and illustrations are part of the McCormick-International Harvester Company Collection, property of The State Historical Society of Wisconsin, and are reproduced with their permission.

Library of Congress Card Number 94-75854

ISBN 1-882256-14-X

94 95 96 97 98 99 00 5 4 3 2 1

Cover and book design by Lou Gordon, Osceola, Wisconsin
Digital imaging by Pixelperfect, Madison, Wisconsin

Printed in the United States of America

PREFACE

The histories of machines and mechanical gadgets are contained in the books, journals, correspondence and personal papers stored in libraries and archives throughout the world. Written in tens of languages, covering thousands of subjects, the stories are recorded in millions of words.

Words are powerful. Yet, the impact of a single image, a photograph or an illustration, often relates more than dozens of pages of text. Fortunately, many of the libraries and archives that house the words also preserve the images.

In the Photo Archive Series, Iconografix reproduces photographs and illustrations selected from public and private collections. The images are chosen to tell a story...to capture the character of their subject. Reproduced as found, they are accompanied by the captions made available by the archive.

The Iconografix Photo Archive Series is dedicated to young and old alike, the enthusiast, the collector and anyone who, like us, is fascinated by "things" mechanical.

The photographs and illustrations that appear in this book were made available by The State Historical Society of Wisconsin. The Society is the official repository for records of the International Harvester Company and its nineteenth-century predecessor, the McCormick Harvesting Machine Company.

The McCormick-International Harvester Company Collection contains nearly 4,000 cubic feet of family papers and business records, including technical publications, advertising literature, engineering and promotional photographs, posters, and films. Its cataloguing, preservation, and administration are funded through an endowment established in 1991 by Brooks McCormick.

Use of the collection is not generally restricted. However, due to its size and complexity, interested persons are encouraged to contact the Society in advance, at 816 State Street, Madison, Wisconsin 53706.

McCORMICK-DEERING
FARMALL

INTRODUCTION

While countless products have influenced the course of modern agriculture, few have had the impact of International Harvester's Farmall. Introduced in 1924, it was the first tractor designed to operate in any type of crop...a true general purpose tractor.

The Farmall evolved from International Harvester's efforts to build a practical motorized cultivator, work which began in 1915. Between 1917 and 1918, IH built 401 motorized cultivators. Their instability, slowness, and high manufacturing cost led the company to suspend production before the 1919 season. However, interest in developing a cultivating tractor and a full range of compatible implements remained intact. Alexander Legge, General Manager of IH, leant his full support to continued experimentation, with the understanding that whatever form the tractor took it "must at least look like a tractor."

A series of prototype tractors and implements were developed beginning in 1920. In 1922, 17 Farmalls were built for extensive field trials. In 1923, 26 of a redesigned, smaller Farmall were built for further evaluation. The 1923 prototype was the basis for the first production run of 200 tractors built in 1924. The McCormick-Deering Farmall was an immediate success, although production was limited until 1926. A total of 117,479 tractors were built through 1930 . The larger Farmall-30 was introduced in 1931, at which time the original Farmall gained the designation of Farmall Regular. It remained in production until 1932, when its successor the Farmall-20 was introduced.

Farmall's distinctive features included: a high-clearance rear axle that permitted the tractor to clear early row-crops for cultivation; 74-inch rear wheel tread width that straddled two rows; and narrow front wheels that ran between two rows. The implements designed for use with the Farmall permitted two or four row operations. The tractor was powered by an International Harvester 4-cylinder vertical engine of 3.25 x 5 inch bore and stroke. Its 3-speed transmission offered forward speeds of 2, 3, and 4 mph. In its Nebraska Test of September 1925, Farmall developed maximum 20.05 belt and 12.7 drawbar horsepower with maximum drawbar pull of 2,727 pounds.

We have included photographs and illustrations of both experimental and production tractors, a good number of which have never been published. With interest in the history of Farmall growing, *Farmall Regular Photo Archive* is certain to be welcomed by all tractor collectors and enthusiasts.

EVOLUTION OF THE FARMALL

A 1917 International Harvester motor cultivator pulling a 7-foot horse disk harrow with tandem attachment. Equipped with a 12 horsepower Le Roi motor, it was the company's first attempt at a universal tractor.

By May 1919, the motor cultivator featured heavier wheels and a strengthened front frame adapted to receive different kinds of attachments. Note the belt pulley.

IH suspended manufacture of motor cultivators in 1919, but had not abandoned the idea of an all-purpose tractor. This March 1920 prototype was a radical step away from the motor cultivator. Power was delivered to the two wide spaced front wheels, and a differential was used for the first time.

Side view of the March 1920 prototype. This was the first machine in which the motor was mounted on the tractor frame with the rear wheel castering independently of the power unit. A Le Roi motor supplied power.

The March 1920 prototype was designed to be driven in either direction. Here, equipped with a planter, the small tiller wheel ran forward.

The March 1920 prototype with two large wheels running forward. IH designed several attachments matched to the new tractor.

The March 1920 prototype with 2-row corn picker. December 1920.

November 1920 prototype with Waukesha engine turned for the first time parallel to the frame. The tractor was reversible, with chain drive, and winch-type power lift. The letter "A" indicates engine crank.

May 1921 prototype, equipped with sweep rake, driven with wide-spaced wheels forward.

May 1921 prototype, equipped with integral cultivator, driven with single front wheel forward.

July 1922 prototype, by which time the tractor was no longer reversible.

February 1923 prototype strongly resembled the production Farmall.

PRODUCTION TRACTORS
1924-1932

Cutaway of the Farmall Regular.

The Farmall assembly line in 1928.

A trainload of Farmalls. July 1928.

Two views of serial no. QC-503, the third Farmall built in 1924 and the first sold.

FARMALL

Farmall Narrow Tread. February 1930.

Farmall with Southern-style open wheels with cone lugs. February 1929.

Rear view of a 1925 Farmall.

Farmall and an experimental 2-way plow. May 1928.

Farmall with an experimental 3-furrow disk plow.

Farmall with power lift middle buster. May 1929.

Farmall with FA-320 planter.

A 1923 Farmall pulling two 2-row corn planters with fertilizer attachments.

Two views of a Farmall and 6-row South American corn drill. June 1926.

A 1925 Farmall with Farmall Drill.

Farmall and two 2-row planters with special hitch. June 1924.

Farmall and 4-row planter. Grinnell, Iowa. June 1927.

Farmall and 4-row side connected planter.
Flood Valley Springs, South Dakota. 1930.

Farmall and two 2-row planters with special hitch. May 1923.

A new McCormick-Deering grain drill behind a Farmall. 1930.

Planting at night with a Farmall and drill.

Farmall with an experimental mounted 4-row corn and cotton planter. October 1928.

Farmall, with wide rim wheels, and drill. Anoka, Minnesota. May 1929.

Farmall and FA-122 4-row planter with fertilizer attachment. Aurora, Illinois.

Early Farmall with front-mounted 2-row cultivator. February 1924.

Farmall, future farmer, and front mounted cultivator. June 1924.

First cultivation with regular 4-row Farmall cultivator. Booneville, Iowa. 1930.

Farmall with integral 2-row cultivator attachment in corn. July 1926.

FARMALL

Front and rear views of a Farmall, with integral cultivator, pulling two cultivator rigs. May 1926.

Farmall and 4-row integral cultivator operating at the IHC Farm, Hinsdale, Illinois. August 1928.

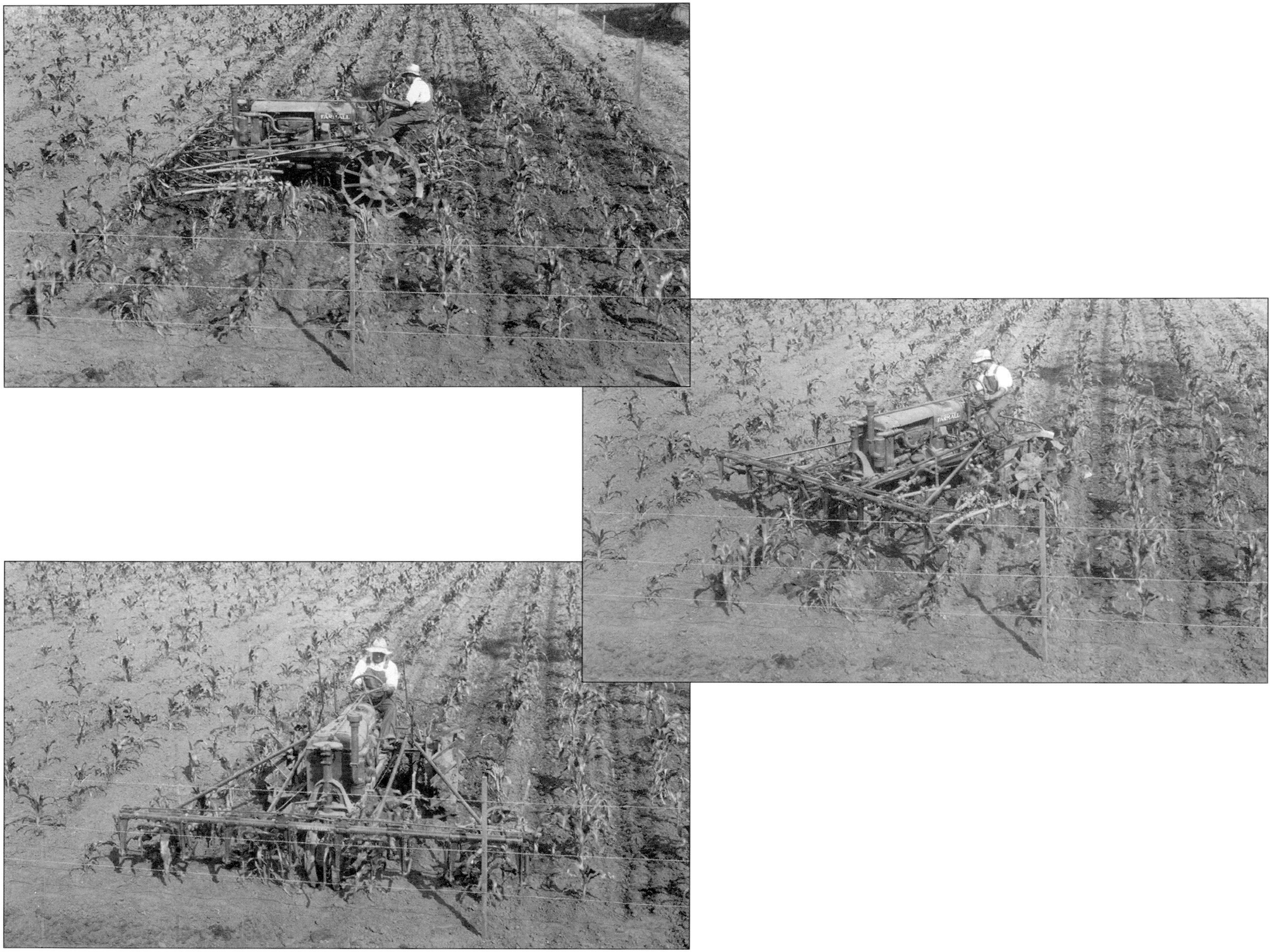

A Farmall with a No. 407-A 4-row cultivator making a turn by backing around. 1930.

Two Farmalls with 2-row integral cultivators. July 1926.

Farmall with experimental front-mounted cultivator. July 1926.

Early Farmall with integral cultivator. 1924.

A 1924 Farmall and mounted cultivator with rotary weeder attachment.

Farmall with the first prototype of the Farmall 4-row integral cultivator. June 1925.

Farmall pulling three rotary hoes. August 1928.

A Presque Isle, Maine farmer and his wife photographed with their new Farmall. November 1926.

Farmall Regular with a No. 2 tractor cultivator. 1935.

Dusting early cotton crop. May 1926.

Spreading lime. October 1927.

FARMALL

Farmall and lime spreader. 1929.

Farmall and mower. July 1925.

A Farmall built in October 1923 and power mower.

Farmall and power mower. 1924.

Farmall with power mower and side-delivery rake. 1926.

Farmall, with optional lighting, and side-delivery rake. September 1924.

Farmall with McCormick-Deering sweep rake. August 1928.

Farmall pulling a wagon and rake bar hay loader. Burlington, Iowa. 1925.

Farmall and a McCormick-Deering hay press. July 1929.

Farmall and McCormick-Deering baler. Libertyville, Illinois. December 1924.

Harvesting rye with a Farmall Regular and an 8-foot McCormick-Deering windrower. Waterman, Illinois. 1931.

McCORMICK-DEERING
WINDROW HARVESTER
FARMALL

The 11 year old son of a Malcom, Iowa farmer operating a Farmall to harvest oats. August 1924.

Farmall with a 10-foot power binder. Des Moines, Iowa. August 1924.

Farmalls cutting lespedza crop. Clarksville, Tennessee. 1930.

Early Farmall and a McCormick-Deering binder. September 1924.

FARMALL

Harvesting wheat with a No. 22 combine and a Farmall Regular with pneumatic tires. Latham, Illinois. August 1935.

Farmall and a McCormick-Deering windrower. Montana State College Farm, Brockton, Montana. August 1928.

Farmall and a No. 20 combine in an Iowa wheat field. Humeston, Iowa. July 1931.

FARMALL

Two Farmall Regulars windrowing on a Sheridan, Wyoming sheep ranch. 1931.

Farmall supplying power to operate a McCormick-Deering thresher. Mt. Morris, Illinois. October 1928.

An early Farmall operating a thresher. Niagwood, Illinois. October 1924.

Farmall with a 4-row bean harvester. Saginaw, Michigan. 1929.

Farmall and a McCormick-Deering cotton picker. August 1927.

Filling a 2-row potato planter. Note fertilizer attachment. Fort Fairfield, Maine. 1930.

Farmall spraying potato plants near Boston, Massachusetts. 1930.

Harvesting potatoes with a McCormick-Deering 2-row potato digger. 1929.

Farmall and a potato digger. Brownsville, Texas. January 1928.

Grading and sacking 1200 bushels of Commercial No. 1 potatoes per day in the field, with home-made outfit powered by a Farmall Regular. Antigo, Wisconsin. 1932.

Farmall and a McCormick-Deering single row potato digger. 1928.

Farmall, 2-row corn picker and 3 wagons. Ashland, Illinois.

Farmall with a McCormick-Deering corn binder and wagon. September 1928.

Early Farmall and McCormick-Deering corn binder, on the IHC Hinsdale farm. November 1923.

Farmall Regular and a McCormick-Deering One-Row Tank Corn Picker. 1931.

Two views of a Farmall and a Power-Drive Corn Picker. Minooka, Illinois. 1929.

McCORMICK DEERING
POWER-DRIVE CORN PICKER
McCORMICK DEERING
FARMALL

Farmall and an experimental 2-row mounted corn picker.

Farmall and pull-type McCormick-Deering corn picker. Altoona, Iowa. 1926.

Farmall and mounted 2-row snapper. Morocco, Indiana. December 1928.

Farmall Regular with an experimental 1-row mounted corn picker with hopper. 1931.

Farmall supplying power to a 6-roll husk shredder. December 1928.

Farmall filling a silo on the IHC Farm. October 1927.

Farmall and a McCormick-Deering power spreader.

Loading a Farmall Power Spreader on a Woodward, Iowa farm. March 1930.

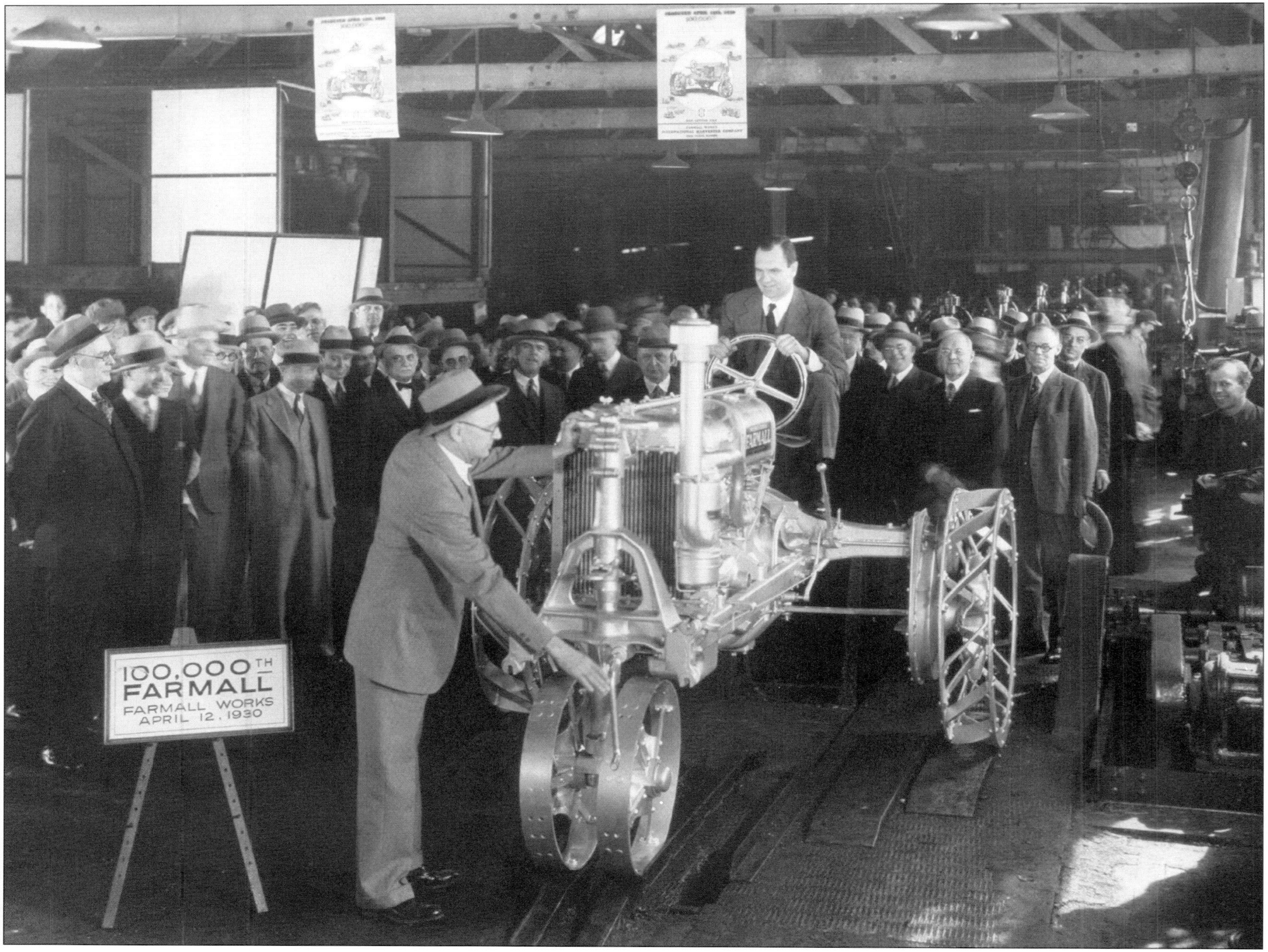
100,000TH
FARMALL
FARMALL WORKS
APRIL 12, 1930

STUDIO ARTWORK

Left hand view of a 1927 Farmall with steel wheels fitted with extension angle grousers.

Right hand view of a 1929 Farmall with steel wheels fitted with extension angle grousers.

McCormick-Deering power mower mounted on a 1927 Farmall.

Farmall and a 2-row cotton and corn planter. 1928.

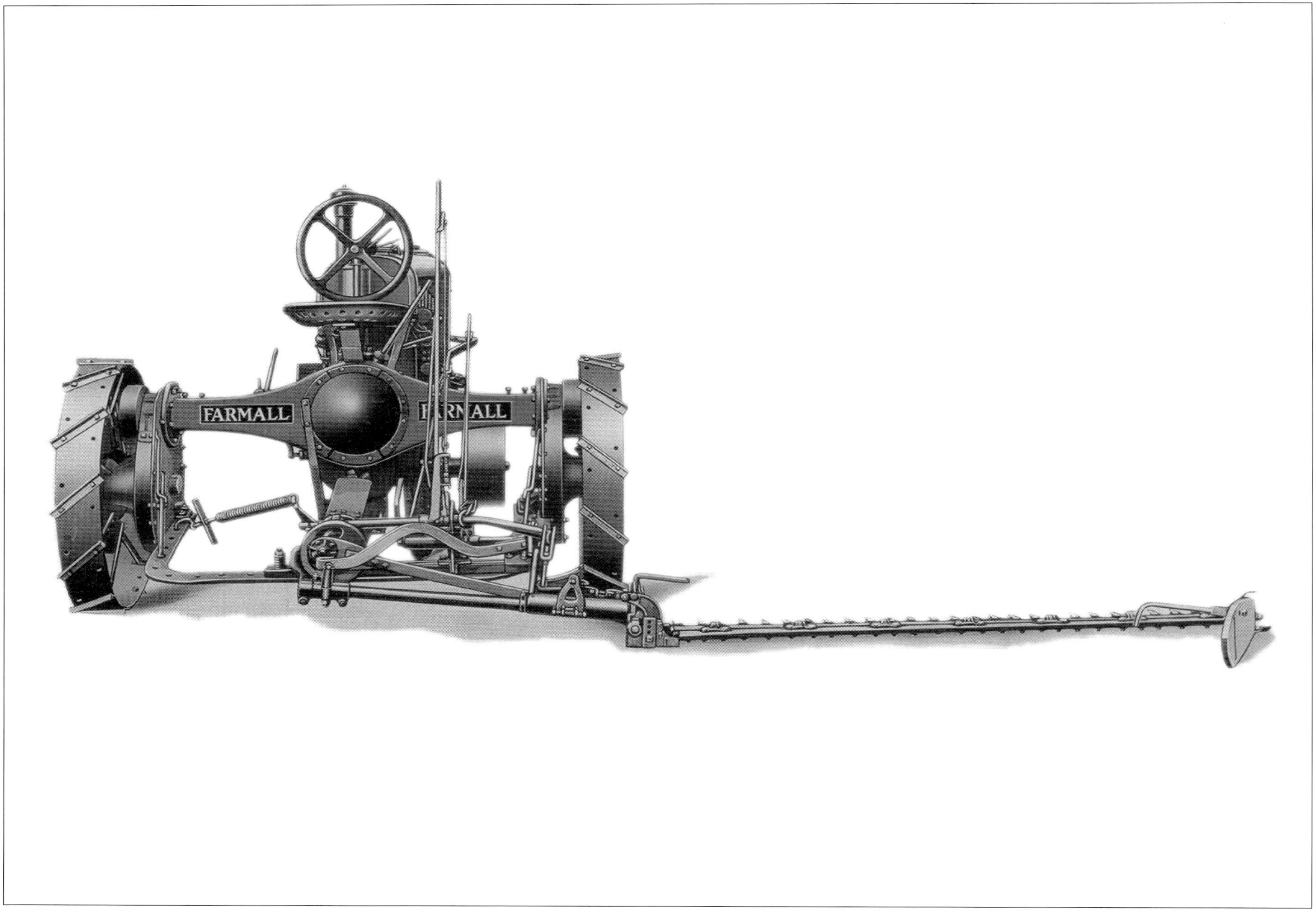

McCormick-Deering power mower mounted on a 1930 Farmall.

Farmall with Purolater oil filter. 1930.

Farmall with mounted power mower and a mechanically driven pull-type mower.

Farmall and a McCormick-Deering One-Row Tank Corn Picker.

Special narrow tread Argentine special. 1930.

Farmall with Purolater oil filter and spade lugs. 1930.

McCormick-International Harvester Negative Numbers

Photographs in this book are part of the McCormick-International Harvester Company Collection, and are kept on file at The State Historical Society of Wisconsin, Visual and Sound Archives, 816 State Street, Madison, Wisconsin 53706. Listed below are the negative numbers or references for these photographs and the pages on which they appear:

Page		**Page**		**Page**		**Page**		**Page**	
8	I-1336-S	38	I-31392		I-1528-T	87	I-28351	113	I-37465
12	I-1600	39	I-31391		I-1530-T	88	I-923-U	114	I-26833
13	GP968	40	I-31649	63	I-1531-T	89	I-926-U	115	I-1744-T
14	GP1372	41	I-33838		I-1529-T	90	I-27805	116	I-1382-S
15	GP1373	42	I-33902	64	I-32129	91	(X3)48889	117	I-1383-S
16	I-21958	43	I-791-T	65	I-31587	92	I-1815-T	118	I-1746-T
17	I-22071		I-793-T	66	I-27823	93	I-1813-T	119	I-28342
18	GP1640	44	I-29229	67	I-27449	94	I-27817	120	I-38071
19	GP1621	45	I-29231	68	GP3665	95	I-27819	121	I-38070
20	GP1893	46	I-475-T	69	I-37196	96	I-1784-Y	122	I-18-U
21	GP1815	47	I-33705	70	I-32612	97	I-37239	123	I-17-U
22	GP2354	48	I-37632	71	I-1584-Y	98	I-1259-U	124	I-38081
23	GP2554	49	I-37633	72	I-31247	99	I-1258-U	125	I-33506
25	I-258-U	50	I-38945	73	I-31248	100	Album 328	126	I-95-T
26	I-36060	51	I-612-U	74	I-4693	101	I-37606	127	I-687-T
27	I-36807	52	I-27247	75	I-34692	102	I-28130	128	Album 440
28	I-1168-JJ	53	I-27248	76	I-1159-S	103	I-599-S	130	I-34437
29	(X3)48890	54	I-27666	77	I-29643	104	I-34374	131	I-38098
30	I-1622-T	55	I-1430-T	78	I-26521	105	I-819-T	132	I-34438
31	344	56	I-31603	79	I-26523	106	I-1236-T	133	I-35351
32	(X3)48892	57	I-31601	80	I-27793	107	I-588-S	134	I-1718-T
33	286/32-1/5	58	I-31150	81	I-31358	108	I-36136	135	I-1714-T
34	I-780-T	59	I-31131	82	I-28004	109	I-1074-W	136	I-1717-T
35	PO10196	60	I-37188	83	I-27707	110	I-37570	137	I-1053-T
36	PO10222	61	I-37186	84	I-37002		I-37569	138	I-1860-T
	PO10220		I-37189	85	I-31687	111	I-37571	139	I-1715-T
37	I-26096	62	I-1527-T	86	I-39275	112	I-1756-T		

The Iconografix Photo Archive Series includes:

JOHN DEERE MODEL D Photo Archive	ISBN 1-882256-00-X
JOHN DEERE MODEL A Photo Archive	ISBN 1-882256-12-3
JOHN DEERE MODEL B Photo Archive	ISBN 1-882256-01-8
JOHN DEERE 30 SERIES Photo Archive	ISBN 1-882256-13-1
FARMALL REGULAR Photo Archive	ISBN 1-882256-14-X
FARMALL F-SERIES Photo Archive	ISBN 1-882256-02-6
FARMALL MODEL H Photo Archive	ISBN 1-882256-03-4
FARMALL MODEL M Photo Archive	ISBN 1-882256-15-8
CATERPILLAR THIRTY Photo Archive	ISBN 1-882256-04-2
CATERPILLAR SIXTY Photo Archive	ISBN 1-882256-05-0
TWIN CITY TRACTOR Photo Archive	ISBN 1-882256-06-9
MINNEAPOLIS-MOLINE U-SERIES Photo Archive	ISBN 1-882256-07-7
HART-PARR Photo Archive	ISBN 1-882256-08-5
OLIVER TRACTOR Photo Archive	ISBN 1-882256-09-3
HOLT TRACTORS Photo Archive	ISBN 1-882256-10-7
RUSSELL GRADERS Photo Archive	ISBN 1-882256-11-5
Available Fall 1994	
CATERPILLAR MILITARY TRACTORS Photo Archive, VOLUME 1	ISBN 1-882256-16-6
CATERPILLAR MILITARY TRACTORS Photo Archive, VOLUME 2	ISBN 1-882256-17-4
MACK MODEL AB Photo Archive	ISBN 1-882256-18-2
MACK MODEL B, 1953-66 Photo Archive	ISBN 1-882256-19-0

The Iconografix Photo Archive Series is available from direct mail specialty book dealers and bookstores throughout the world, or can be ordered from the publisher.

For information write to:
Iconografix
P.O. Box 609
Osceola, Wisconsin 54020 USA

Telephone: (715) 294-2792
(800) 289-3504 (USA and Canada)
Fax: (715) 294-3414